Did you enjoy this issue of BioCoder?

Sign up and we'll deliver future issues and news about the community for FREE.

http://oreilly.com/go/biocoder-news

BioCoder #12

MAY 2017

Beijing · Boston · Farnham · Sebastopol · Tokyo   O'REILLY®

BioCoder #12

Copyright © 2017 O'Reilly Media, Inc. All rights reserved.

Printed in the United States of America.

Published by O'Reilly Media, Inc., 1005 Gravenstein Highway North, Sebastopol, CA 95472.

O'Reilly books may be purchased for educational, business, or sales promotional use. Online editions are also available for most titles (*http://oreilly.com/safari*). For more information, contact our corporate/institutional sales department: 800-998-9938 or *corporate@oreilly.com*.

Editors: Mike Loukides and Nina DiPrimio	**Interior Designer:** David Futato
Production Editors: Colleen Lobner and Nicholas Adams	**Cover Designer:** Randy Comer
	Illustrator: Rebecca Demarest
Copyeditor: Amanda Kersey	

May 2017:   First Edition

Revision History for the First Edition

2017-04-26:   First Release

978-1-491-97654-8

[LSI]

Contents

Welcome to BioCoder

Mike Loukides

It's been an exciting few months for biologists. The first stage of the battle for CRISPR-Cas9 patents (*http://tcrn.ch/2paI8aP*) has come to an end, though it will certainly continue in the appeals courts, and in courts outside the US. How much of an effect does this decision have? I don't know. When I wear my cynic's hat, all it does is change whether entrepreneurs pay rent to Harvard or Berkeley. But conflict and uncertainty are never helpful, and the end of that conflict is now one step closer to being in sight. New York's Genspace (*https://www.genspace.org/*) has had several hundred students attending workshops and classes about CRISPR, and I'd assume that other community labs have seen similar interest. It would be tragic if all that creativity went down the drain in a swirl of legal fees and patent litigation.

More recently, we've read about the Sc2.0 (*http://syntheticyeast.org/*) project. Sc2.0 is synthesizing a complete yeast genome: at this point, they have synthesized six out of a total of 16 chromosomes, and installed them in functioning cells. The yeast genome seems to be very tolerant to editing: you can evidently change a lot of things without harming the yeast itself. It looks like they will have synthesized the complete genome by the end of 2017.

Yeasts are the basis for fermentation; bakers have been using this particular yeast in bread for thousands of years. If programmed correctly, yeasts can make much more than bread and beer. We're already using yeasts to manufacture vaccines, fragrances and flavors (*http://bit.ly/2oiwmHp*), and many other commodities. The possibilities explode when you have a yeast whose genome you can control. They get even greater when you add a new, 17th chromosome (*http://wapo.st/2oERdFH*) that's purely for experimentation. This could be a new standard platform for synthetic biology, the biological chassis that's needed to get results that are much more reliable and reproducible.

As always, we're looking for your ideas and articles. We're interested in almost anything in biology, bioengineering, synthetic biology, DIY biology, open source biology, and related fields. That includes neuroscience, health, and a lot of areas that we don't yet have names for. Here's a specific challenge: what are you doing with CRISPR? What are you doing with yeast? Send your ideas about those topics—or anything else that interests you—to us at *biocoder@oreilly.com*

Citizen Salmon

Glen Martin

Seattle is rich in cultural icons. Think artisanal coffee, grunge rock, the Space Needle, Pike Place Market: they are all instantly recognizable symbols of the city.

But Seattle is more than a city. As the foremost urban center of the Pacific Northwest—or Cascadia, in the preferred regional parlance—its significance and influence are broad. Seattle's symbols are thus Cascadia's symbols. And nothing is more emblematic of Cascadia than salmon. The five species of Pacific salmon—chinook, coho, pink, sockeye, and chum—have both economic and emotional significance to the people of the Pacific Northwest. Residents enjoy catching them and eating them, but the fish also serve as indicators for the health of the regional biome. Moreover, they are totems. Cascadia people don't just value salmon. They revere them.

So the whole issue of sustainable seafood—especially sustainably harvested salmon—has particular resonance in Seattle. Not only are farmed salmon spurned, but locals want to know the source of their wild salmon: who caught it, where it was caught, and how it was caught. But there's no way to really confirm that. A salmon steak or fillet, after all, doesn't come with a resume.

That might change if a Seattle-based group of salmonid-loving biohackers have their way. They've launched Citizen Salmon (*http://citizensalmon.org*), a DIY initiative aimed at genotyping Pacific salmon. The object of the project is to identify a wide range of single nucleotide polymorphisms (SNPS) in the salmon genome. Salmon that hatch in (and ultimately return to) one stream may diverge genetically from salmon that spawn in a stream only a few miles distant. Such variation can allow scientists to differentiate the fish to a rather exquisite degree.

Ultimately, Citizen Salmon could aid fisheries agencies in managing harvest regulations. A major problem in establishing salmon quotas is the difficulty in connecting salmon at sea with the specific natal rivers they return to for spawning.

For example, say River X has a healthy run of salmon, but Stream Y doesn't, and scientists want to do everything possible to bump up Stream Y's populations. Figuring out which individuals in a group of salmon in the ocean are Stream Y fish is daunting unless you have access to a broad genetic database. By confirming that certain SNPs are unique to Stream Y fish, researchers can calculate the number of Stream Y fish among any group of salmon foraging in a given location. If Stream Y fish are particularly abundant, regulators can close fishing for that locale, assuring enough salmon will make it back to Stream Y to sustain, or even increase, the run.

Conceivably, such information could also be used to help consumers source their salmon. By sending in samples to a lab for analysis, fish lovers can determine not only the species of the salmon they're eating, but the origins; they'll be able to tell if their Atlantic salmon was raised in pens in a Norwegian fjord or in a Scottish loch, or if that chinook salmon steak came from a wild fish that spawned in Alaska's Copper River, or from a Sacramento River chinook that was reared from an egg to fingerling size in a hatchery, released to the river, and matured in the ocean prior to returning to spawn. In short, environmentally aware seafood aficionados could judge whether the salmon on their plate conforms to their own metrics for sustainability.

"Not that genotyping will prove absolutely foolproof in linking every fish we sample to its stream of origin," says Michal Galdzicki, Citizen Salmon's founder and prime mover. "There are going to be variations among individual fish that could result in some faulty conclusions, and besides, not all salmon necessarily return to their natal streams. They can be opportunistic. They can go up other streams if it suits their purposes. But on the whole, the techniques we're developing could give us a pretty clear and accurate picture on salmon origins and populations." While Galdzicki is enthused that Citizen Salmon's work might yield effective tools for salmon conservation, he acknowledges that wasn't the foremost motivation for launching the initiative.

"I love biotechnology," he says, "so for me, this was a mainly a matter of using a grassroots project as a means for teaching biotech to the local community. Because we're dealing with salmon, the response was incredible. Around here, you're going to get a lot of attention and support if you're involved in anything that could help sustain and improve salmon populations."

Galdzicki, co-directors Regina Wu and Zach Mueller, and a cadre of volunteers utilize restriction enzyme digestion and direct sequencing to ID Pacific salmon. "We do the restriction enzyme digests ourselves, and we use a commercial lab for the direct sequencing to verify our work," Galdzicki says. "Right now, we're concentrating on chinook salmon, mainly because of the large body of chinook data that already exists. But as our work proceeds and our database expands, we hope to address the other four species of Pacific salmon."

While identifying subspecies, races, and runs of salmon via genetic sequencing is not new, Citizen Salmon represents an opportunity for both refining and ramping up the techniques, says Galdzicki. The initiative also points the way toward enhanced collaboration between citizens and regulatory agencies, a development that could only benefit North America's threatened fish. Citizen Salmon recently joined the Klondike Creek Alliance, an ad hoc group focused on restoring a local watershed, and Galdzicki and his cohorts are engaged in outreach to other conservation and community groups.

"Currently, people here are often engaged with salmon from the cultural and policy ends," says Galdzicki. "Or from the consumer end—a lot of people fish for salmon. But we want to get them better grounded in the science. That could increase their appreciation for salmon and also make them more effective as advocates. I love the idea of someone catching a salmon or two, then genotyping them and contributing to the Citizen Salmon database. That kind of engagement really contributes to long-term and effective stewardship."

Ultimately, says Galdzicki, he'd like to develop standardized tests that will let people quickly evaluate salmon genetics in their own homes.

"We have some promising ideas," Galdzicki says. "The equipment and the techniques are becoming robust and cheap enough that I think we'll see the day when citizen scientists can do this work quickly and accurately in their own kitchens. Our goal is to scale this technology up and get it to as many people as possible. The more we know about salmon and their habitats, the better our chances for conserving them."

Glen Martin covered science and the environment for the San Francisco Chronicle for 17 years. His work has appeared in more than 50 magazines and online outlets, including Discover, Science Digest, Wired, The Utne Reader, The Huffington Post, National Wildlife, Audubon, Outside, Men's Journal, and Reader's Digest. His latest book, Game Changer: Animal Rights and the Fate of Africa's Wildlife, was published in 2012 by the University of California Press.

Defining Open: BioHack the Planet

Wakanene Sebastian Kamau

How do we define an open community and what would we want from one? The inaugural BioHack the Planet conference took place this September in Oakland, California. Hosted in Omni Commons, the volunteer collective home to the DIY-bio hub Counter Culture Labs (*https://counterculturelabs.org/*), the conference embodied the spirit of the community it sought to bring together. BioHack the Planet (BioHTP) was developed to be "run by BioHackers, designed for BioHackers, with talks solicited from BioHackers." Its two organizers, Karen Ingram and Josiah Zayner, are both veterans of the do-it-yourself (DIY) biology, or biohacking community, in their own right. Ingram is a coauthor of the BioBuilder curriculum for teaching synthetic biology. Zayner's DIY supply store, ODIN, has outfitted many budding scientists with everything from pipette tips and bulk cell media to at-home CRISPR kits. So what happens when biohackers organize themselves to discuss their work? Part symposium, part workshop, and part exhibition, BioHTP explored the projects the biohacking community is tackling, how the community is organized, and where it's going.

While BioHTP isn't the first time the biohacking community has been brought together, it may be the first time it's done so by its own at this scale. Soon after the launch of DIYbio.org in 2008, the FBI co-sponsored the "Building Bridges Around Building Genomes" conference in collaboration with the American Association For The Advancement Of Science (AAAS), the U.S. Department of Health and Human Services, and the Department of State to encourage communication between policymakers and academics as well as synthetic biologists working in industry and in community labs. Later, in 2012, the FBI flew in biohackers from around the world to a DIYbio conference to develop their relations with the community. Since then, DIY biologists have organized largely on a continental basis, as in the formation of a European network of do-it-yourself biology, on the web, as in the Hackteria platform, or complementary to larger hobbyist conferences, such in the Nation of Makers Initiative or South by Southwest.

It was after South by Southwest, the annual music, film, and interactive media festival in Austin, Texas, where, in 2014, Ingram and Zayner took part in an impromptu DIY biology panel that they began to envision what a biohacking conference would look like.

"At the moment there is a big hole when it comes to connecting these people [biohackers] who live all across the world," says Zayner. "Most of us are loosely connected on Facebook and Twitter but still many people don't know each other or what everyone is working on." Without these linkages, the community suffers. As Zayner went on to say, "In science, it is extremely important to give and receive feedback on our research, be inspired by each other's work, and collaborate."

Collaborations at Scale

A recurring theme throughout the conference was a call to collaborate. Ranging from the pharmacological, with discussions on the Open Insulin and Open Estrogen projects led by Anthony Di Franco and Mary Tsang, respectively, to the ecological, with presentation of citizen science efforts like the Oxalis Genome Project by Sebastian Cocioba and Citizen Salmon project by Michal Galdzicki, the DIYbio community has a healthy sense of teamwork.

Part of the motivation behind the DIYbio community push toward collaborative projects is to use them as teaching tools. Take, for example, the Oxalis Genome Project, an ambitious effort to develop the common shamrock-looking plant into a model organism. As Cocioba says, "We want to give anyone and everyone a chance at contributing to the body of knowledge about this plant as an introduction to DNA sequencing, bioinformatics, ecology, etc." By creating public and open repositories for the genomic markers and protocols, Cocibioa hopes to encourage broad participation in the project.

Other projects were more explicitly directed at developing biological tools for scientists. Kate Adamala presented work on building a synthetic cell from scratch to determine the requirements for life in a manner similar to breadboards in the study of electronics, while Keoni Gandall's efforts to develop a modular genome from the bacterium Mesoplasma florum. Both showcased efforts to develop platforms for prototyping biology. In Gandall's case, he wants to make his work on the bacterium "open, so anyone can use it."

A number of presenters used the conference as an opportunity to crowdsource suggestions to overcome hurdles. If two heads are better than one, then perhaps a crowd is better than two? This was certainly the case for Citizen Salmon (*https://www.oreilly.com/ideas/biohacking-with-citizen-salmon*), a DNA barcoding effort to trace store-bought salmon to the lakes and streams it comes from. "We have a particularly big challenge," says project lead Galdzicki. "Most of the organisms people work with today already have genomes sequenced. There is no [public] genome available for the Pacific salmon." The following Q&A session provided a spirited discussion of novel genome sequencing methods and enlivened the group's efforts.

Moreover, as Cory Tobin, from The L4B, showed with his presentation on *Streptomyces thermoautotrophicus*, a bacteria purported to be able convert nitrogen from the atmosphere into ammonia in the presence of oxygen, collaborations are can help push a project to completion. By recording his early work, as he would in a lab notebook, on a public wiki page, he was able to catch the eye of a number of academic researchers. Together, they formed a multi-institution consortium to study the bacterium. Support from the consortium leveraged the resources to sequence and align *S.thermoautotrophicus'* genome, which ultimately led them to conclude that the bacterium does not fix nitrogen, contrary to what earlier studies had reported. Slightly dismayed in their final result but confident in their work, Tobin urged researchers to publish their results. "Someone else probably already tried this, he said, "and it didn't work." A message he and his collaborators took to heart by publishing their negative results (*http://www.nature.com/articles/srep20086*) in the open access scientific journal *Scientific Reports*.

Transdisciplinary Cooperation

Not only have collaborations flourished within the traditional science community as the open science movement has grown, but, as BioHTP shows, open collaboration has given nontraditional and traditional scientists the ability to biohack.

Representing the group of designers and landscape architects behind the collective BkBioReactor, biologist Elizabeth Hénaff presented on an investigation of the Gowanus Canal (*https://www.oreilly.com/ideas/the-bk-bioreactor-takes-on-the-gowanus-canal*), a Superfund site in Brooklyn. As an artist-in-residence at the School of Visual Arts Visual Future Lab, Hénaff designed a series of bioreactors, inspired by the architecture of sponges, to host bacteria that would promote a remediation plan based on microbes. Reflecting on her residency and the project, Hénaff came to the realization that "[Design] is such a rich field that has been studying how to communicate things. That is something that, as scientists, we do see the need of doing, or don't do as well." As the project continues to evolve, Hénaff proposed expanding the project to include other Superfund sites. "For better or for worse, I think everybody probably has a Superfund site in their back-yard," she says. "What I would like to see is if we could have a collective effort to characterize the the microbiome of all these different superfund sites and see if we can come up with a solution together." By sharing the results of culture experiments, process and design files, and adopting a distributed sequencing effort, similar to the Oxalis Genome Project, the project develops an open network of DIY spaces studying the environmental metagenomics of Superfund sites.

Beer brewer Nick Moench and dog breeder David Ishee showed how the availability and accessibility of biology protocols can lead nontraditional scientists to new DIYbio applications. "Craft beer has a problem," Moench says. "We all use the same microbes to make the same beers from the same labs and it's boring." Expanding the biodiversity of microbes used by his company, Inoculum Ale Works, to create its unique sour beers drives his passion. For Moench, soliciting microbes from the biohacking community is a way to develop both his beer and the network of people changing the course of the beer brewing industry. As he says, half tongue-in-cheek, "If you can't be the best, at least be the coolest."

For Ishee of Midgard Mastiffs (*http://midgardmastiffs.com/*), the goal is to use techniques from genetic engineering to breed dogs with fewer genetic disea-ses. Ishee, who does not have a formal background in biology, saw the unrealized potential for biohacking in this field. "Now, we have the technology to solve it, but big institutions have no interest in it," he says. "There's no profit motive. Dog breeders would do it, but we don't have the technology or the know-how. But that's exactly what I am trying to do: bridge the gap between dog breeder and genetic engineer." While his first attempt at using sperm-mediated gene transfer, a technique that uses sperm to transfer exogenous genetic information during fer-tilization, to confer green fluorescence to dogs was unsuccessful, he remains opti-mistic and committed to learning and experimenting. "I'll try again, with a few changes, to increase the chances of fertilization."

Responsible Openness

With good reason, many of the projects at BioHTP emphasized and encouraged collaboration. As we've seen, collaboration can be insightful, instructive, and facil-itative, bringing in energy or inspiring confidence in early-stage projects. Perhaps now, when collaboration has become a cultural standard and goes seemingly hand in hand with the term "open," we can begin to have a dialogue on what kinds of collaboration should be encouraged and what responsibilities are taken on by open networks. From the conference, four speakers were particularly interested in discussing the structure of the biohacking community: David Kong, Thomas Landrain, Megan Palmer, and Drew Endy.

David Kong, cofounder of the international How to Grow (Almost) Anything course on synthetic biology, touched on the requirements for developing a "Creative Ecosystem of Making." He cited a finding from a study by Jeppesen and Lakhani in the journal *Organizational Science* that showed that participants who were in the "outer circle" of the scientific establishment" were able to perform with greater success than those from within the establishment. These outliers could capitalize on creativity to drive innovation. To foster this creativity, Kong highlighted a need to "make 'biological making' radically diverse." In order to do so, his talk focused on the tools, the spaces, and the communities that make up an ecosystem of creative making. From his background in microfluidics, a study of the small-scale manipulation of fluids which has wide application in synthetic biology, Kong has been part of a team that is developing an open, online repository for the designs of printable microfluidic devices. Additionally, Kong cofounded the cross-cultural community center, East Meets West, that hosted the How to Grow (Almost) Anything course in 2015. The center also serves as a bookstore, art gallery, and performance venue for the greater Cambridge community.

In thinking the concept of community on a larger scale, Thomas Landrain used the metaphor of dark matter to explain why he was driven to help create the French biohackerspace La Paillasse in 2011:

> Dark matter and dark energy comprises 95% of the universe, and it's stuff you can't interact with. [...] As a researcher, at some point I realized that we were such a small minority that actually had the monopoly of doing science somehow. Why couldn't we imagine ways to bring in more people? People outside academia are trying new ways of producing science, but that couldn't happen within traditional matter, academia, so we had to try to interact with that dark matter.

It was in this spirit that, in 2015, La Paillasse hosted a six-month, online, open cancer epidemiology initiative with the pharmaceutical giant Roche. The result, Epidemium, included over 300 participants from across Paris and yielded over 3,600 contributions to their public wiki. Notably, the initiative cut across disciplines. Of the participants, one-third were students, and two-thirds were professionals, many of whom worked outside of health industry. With an eye to collaboration on an even larger scale, Landrain is now focused on understanding the social science behind how open science groups collaborate through the think tank CommonGround.

In parallel, Megan Palmer's presentation, "BioHacking Governance" opened a dialogue for discussing the ethos of "doing it yourself" within the biohacking community and showcased ways that the biohacking community has regulated itself. As Palmer noted, within the biohacking community, there is often a lack of specificity about its values and what systems it supports. For example, she says, "We talk about democratizing technologies, but what does that mean? Does that mean just access to tools, or does that mean equity on how those tools are used and the results of it?" For Palmer, questions like these do the work of teasing out the underlying sentiments conveyed by the popular contrarian narrative within the biohacking community. Nonetheless, methodologies in this vein are not without their critics. As an attendee noted during the Q&A, "When you have a community that has community standards and norms and does communicate, you do in fact create [a] system, create ways of collaboration, and create rules as far as safety." A valid claim, which, on a closer look, is less at odds with the message of the presentation than it appears. By beginning to shift away from mentalities of "doing it yourself" to new narratives of "doing it together" we can finally ask, to quote Palmer, "What are the principles of systems we want to encode?"

Outlining these principles is hard work. It requires negotiating logistics and interests across a wide range of interested parties. With due consideration, the entirety of the last day of the three-day conference was devoted to a series of roundtable style community discussions on topics ranging from biosafety and funding to collaborations and plans for future conferences. Participation in these conversations is crucially important and, sadly, was not as strong as it ought to have been, given the turnout for the rest of the conference. Geographies notwithstanding, only one person joined in online. By distributing power and responsibility across our network through conversations like these, we empower the entire biohacking community. By cultivating a culture of participation at community discussions, we can identify what our community's cultures and norms are and do the work of encoding a system of principles that works for us. "The goal of biohacking is to make science accessible," Zayner says. "Most of us take that seriously and go out of our way to write protocols or share materials, but we need to do much more than that. We also need to go out of our way to make science available by welcoming those different from us through encouragement and direct outreach." Zayner and Ingram say that to foster an inclusive environment, the movement must encourage representation across different aspects of identity: gender, ethnicity, race, education level, age, expertise, desired outcomes for the technology, income, ability, and so on. Active participation in community-oriented dialogue is a civic responsibility to ourselves, and we need to do better.

Conceptualizing what citizenship might mean in the context of biohacking was the focus of the keynote presentation by synthetic biology pioneer Drew Endy. He used an anecdote about his attempt, and subsequent failure (*http://biorxiv.org/content/early/2015/08/13/024299*), to produce a batch of the homegrown opiates that raised public concerns over DIY biology to introduce the following quote (*http://bit.ly/20ERA2S*) from Thomas Jefferson to John Adams: "A government adapted [...] for the Man of these states. Here every one may have land to labor for himself if he chuses [*sic*]; or, preferring the exercise of any other industry, may exact for it such compensation as not only to afford a comfortable subsistence [.]"

For Endy, this gets to to why biohacking is important: it is "a path by which people can secure a capacity for making things, biological and otherwise."

To solidify this path, Endy has been active in developing tools that broaden access and increase the interconnectedness of biological labs.The newest tool, the Bionet, is a project to coordinate the physical transfer of materials across people doing biology. This includes setting up material nodes at both academic and community labs as well as providing freely available schematics for DIY material scanners and low-cost commercial alternatives. The Bionet, in conjunction with the Open Materials Transfer Agreement (OpenMTA (*https://biobricks.org/open mta/*)), facilities easier redistribution of biological materials. Plans for the Bionet include provisions for crowdsourced participation in deciding what should be made freely available across the system and possibly using geographic need to distribute materials where they are needed most.

Broadly, these tools are allowing for a new cultural tradition within synthetic biology. As Endy emphasizes, our intense focus on application and danger have stunted the growth of the culture surrounding biotechnology. As the barriers around technology for biology are lowered, we as a biohacking community can take an active role in positing toward a future of increased access, with an aim for broader representation.

To echo Endy's parting words, "We can connect with play and design and art by hacking and creating and distributing and enabling people to be citizens by giving the means of production around biohacking." By broadening representation, we can empower citizens to begin to ask what would be meaningful to them in the context of who they are and where they are. BioHack The Planet was a display of what has brought meaning to those who have had the ability to create. As the tools and cultures surrounding openness develop, we have a responsibility to shape how it matures. I hope we take that responsibility seriously.

NOTE
Other incredible presenters at BioHTP whose work was not included here: Heather Dewey-Hagborg (*http://www.dewey hagborg.com/*), Eri Gentry (*http://www.iftf.org/erigentry/*), Structure Films (*http://structurefilms.com/*), David Brown (*https://www.linkedin.com/in/david-brown-0830a8b0*), and Kate McLean and Mario Furloni (*https://vimeo.com/ 27023498*).

Wakanene Sebastian Kamau is a designer and writer based in Seattle. His work has appeared in CLOT Magazine, The Future of Life, Science and Society Review, and Scientia. He is affili-ated with the biohackerspace SoundBio Lab (http://sound.bio) and holds a BSc in biological chemistry from the University of Chicago. You can follow him on Twitter @ws_kamau (https://twitter.com/ws_kamau).

BK BioReactor

Glen Martin

Brooklyn is perhaps New York City's most vibrant borough, drawing the young, the visionary, and the entrepreneurial. Its restaurants express the cutting edge of culinary theory, and it's a hotspot for IT and biotech start-ups. Maybe you still go to Manhattan to make money, but you go to Brooklyn to make a difference (and also some money, of course).

But it's not all high-end brewpubs and succulent tapas for Brooklyn. The borough has some problems, many of them the legacy of long human habitation. Case in point: the Gowanus Canal.

Originally a network of tidal creeks and wetlands rich in fish and wildlife, the canal was transformed into a shipping hub in the 19th century, evolving into the nation's busiest commercial channel in the years following World War 1. But such intensive development exacted a price. Decades of industrial discharges, sewage outflows, and contaminated runoff made the canal one of the country's most polluted waterways by the end of the last century.

Today, the 1.8-mile canal has lost its industrial and maritime preeminence. The dire legacy of its commercial glory years, however, endures. While it still serves as a conduit for the transportation of fossil fuels and scrap metal, the canal is most notable—or notorious—as a Superfund site. Clean-up efforts are underway, but the water in the canal remains a toxic, fetid brew laced with heavy metals, raw sewage, and polycyclic aromatic hydrocarbons. Dissolved oxygen levels are a mere 1.5 parts per million (ppm), far below the 4 ppm minimum required to sustain life.

Actually, that last statement should be amended. Yes, the water in the Gowanus Canal is too polluted to support aquatic life as most people think of it. Any fish that moseys in from the East River will go fins up. It can't even sustain oysters, which are pretty tough mollusks. But the canal still teems with life—microbial life.

"The bottom of the canal is covered by 10 to 15 feet of viscous, black, tar-like sediment," says Elizabeth Hénaff, a New York bioengineer. "It's an extremely rich environment for microorganisms. When we take a close look, we find there are two different environments that favor two broad categories of microbes. The first requires an anaerobic marine environment, and the second thrives in the human gut—due to sewage overflows, portions of the canal support robust communities of microbes that typically reside in the human intestinal tract."

From a microbiologist's point of view, then, the Gowanus Canal is an analogue of the Amazon or the Mariana Trench: a still largely unexplored biome teeming with organisms that beg for cataloging and study. And Hénaff and her colleagues—Matthew Seibert, Ian Quate, Ellen Jorgensen, Christopher Mason, and Benjamin Wellington—are up to the task. They've formed BK BioReactor (*http://www.bkbioreactor.com/*), a collaboration designed, as the partners put it, to "investigate Brooklyn's hippest Superfund."

Hénaff observes the project was spurred by a US Environmental Protection Agency plan to dredge and cap the waterway over the coming decade. While the work is necessary to mitigate the public health and environmental threats posed by the contamination, such remediation will also obliterate the canal's singular—albeit artificial and toxic—ecosystem.

"We're discovering that these highly contaminated areas support microbial communities that are unique," Seibert says. "They evolve in extremely polluted urbanized environments—in the case of the Gowanus Canal, over a period of 150 years. That fact alone qualifies them for taxonomic research, but they could also prove a source for cell products. We need to find out what's there before they're destroyed."

The value of microorganisms that have evolved in an environment as contaminated as Gowanus is obvious to any microbiologist. Many of the microbes don't merely exist with toxic elements and compounds; they subsist off them. That makes them promising agents for bioremediation efforts—though some tinkering inevitably will be required to maximize their potential.

"Their metabolic functions allow them to degrade toxic compounds, but they generally don't do it fast enough for human time scales," says Quate. "So to accelerate their desired functions, we can approach it from two ways. We can modify the bacteria to overexpress the relevant pathways, or we can modify the bacteria's environment to favor the growth of the microbial communities we want." The first option is suitable for treating contaminated material in a controlled environment, says Hénaff.

"In the case of Gowanus, you'd dredge toxic material, isolate it, and treat it," she says. "Obviously, you couldn't release modified bacteria into the environment."

The second strategy consists of creating built environments that encourage the growth of pollutant-degrading microbes. Hénaff observes that various substrates and shapes in nature encourage different microbial communities. "Sea sponges and corals, for example, live in symbiosis with whole hosts of bacteria," she says. "The little niches and cavities in corals and sponges create microenvironments where you get different pHs, temperatures, and water and nutrient flows. So we're creating microbioreactors that allow us to experiment with canal sediment and water and different types of materials in different configurations. The goals are to identify the communities that foster the (toxic compound) degradations we want, and then nudge them toward greater efficiency."

The researchers acknowledge the parameters of "greater efficiency" are still undefined. It's possible that the degradation of toxic compounds could be accelerated rapidly, either by modified bacteria in isolated areas or through in situ strategies that focus on the manipulation of structures and materials in the built environment. On the other hand—well, it may prove difficult to improve on nature to any significant degree.

"Maybe we'll find we can accelerate degradation from thousands (of microbial) generations to two," says Seibert. "Or maybe it'll only be a reduction to hundreds of generations. Or even less. We won't know until we do the research. But this is a rich area for both microbial modification and urban design, and it needs to be pursued."

Ultimately, the partners see BK BioReactor growing into a comprehensive research library for Superfund microbes. At Gowanus, the team hopes to install "smart docks" that can sample and evaluate the rich microbiomes teeming in the canal's sediment; they also plan to outfit a boat that could serve as a mobile "branch" of the library.

"Ultimately, we could take it to other Superfund sites," says Quate. "They all represent unknown territory, and that's tremendously exciting. There's no telling what we'll find there."

Glen Martin covered science and the environment for the San Francisco Chronicle for 17 years. His work has appeared in more than 50 magazines and online outlets, including Discover, Science Digest, Wired, The Utne Reader, The Huffington Post, National Wildlife, Audubon, Outside, Men's Journal, and Reader's Digest. His latest book, Game Changer: Animal Rights and the Fate of Africa's Wildlife, was published in 2012 by the University of California Press.

Amazing, Super-Sweet Natural Proteins

Raj Nagarajan, PhD

Many of us have a sweet tooth and crave sweet treats throughout the day, like chocolate, candies, fudge, ice cream—you name it. Some of us consume these a little above the recommended amounts because sometimes sweet food just makes us feel better.

Sucrose is the major carbohydrate that is being consumed in every form of processed food such as confectionery, dairy products, or soft drinks. A food company may add honey (40% fructose and 35% glucose), date sugar (80% sucrose), rice syrup (100% glucose), corn syrup (98% glucose), coconut sugar (70%–80% glucose), or evaporated cane sugar (sucrose). But those are just sugar in different forms.

Many processed food products thus contain free sugar, which is concentrated form of sugar. The problem with free sugars is that they do not go through rigorous digestion because they are already released or stripped out of their original encasement (e.g., plant fiber). Free sugar rushes into the small intestine, where sugar metabolism occurs. If the sugar is sucrose, it undergoes hydrolysis into glucose and fructose. Freed-up glucose and fructose travel through the hepatic portal system (which contains the veins and its tributaries from the gastrointestinal tract to the rectum) and get into the bloodstream and flow through the liver for additional processing; fructose is processed into glucose-like compounds for ATP production (energy molecule), while glucose serves as a substrate for the Krebs cycle (cellular respiration) for ATP production. Surplus glucose is converted into glycogen for long-term storage in the liver. If the glycogen accumulates rapidly due to consumption of excess sugary food, it can then undergo a metabolic process that transforms glycogen into fat or LDL cholesterol (the bad cholesterol).

Is there a way around overdoing it? Instead of avoiding or regulating consumption, many people try limiting their sugar intake with artificial sweeteners—

sugar substitutes such as saccharin, aspartame, and splenda (sucralose) that taste like sugar but have zero or few calories.

Aspartame, the most popular synthetic sweetener, is made by combining two amino acids: aspartic acid and phenylalanine. But from these components, you would not know that aspartame is 200 times sweeter than sugar. Phenylalanine tastes bitter, and aspartic acid has a flat taste. The chemical combination modifies phenylalanine, resulting in sweet taste perception. Larger compounds can be formed from two simple molecules with different properties from their unlinked components. Although aspartame consists of amino acids, it is not a protein but a dipeptide, which is made up of two amino acids.

Aspartame and other synthetic sweeteners are approved as safe for consumption by the Food and Drug Administration. Numerous research investigations have also found no correlation between aspartame consumptions and brain tumors, diabetes, seizures, and childhood behavioral disorders. More than 200 scientific studies have confirmed that aspartame is a safe sugar substitute. Beyond this, American Medical Association, the American Dietetic Association, the American Diabetes Association, the World Health Organization, the European Commission, and the food regulatory agencies of over 100 countries have approved aspartame as a safe sugar substitute.

In the midst of very positive research results, scientists have also discovered worrisome side effects. For example, John Olney, a well-known psychiatrist and neuropathologist at Washington University School of Medicine, and Richard Wurtman, emeritus professor of neuropharmacology at the Massachusetts Institute of Technology, revealed a link between aspartame and altered brain chemistry that may lead to brain tumors and seizures. A 2014 study by Ohio University researchers strongly linked cardiovascular diseases to diet drinks consumption in post-menopausal women.

What could be responsible for these adverse effects of artificial sweeteners? Critics hypothesize that the methyl group (CH_3) attached to the phenylalanine amino acid in aspartame is unstable at 30° C and releases methanol (CH_3OH), which is toxic and is linked to causing blindness. Excess methanol in the blood can reach the liver and later can be converted to formaldehyde (CH_2O), a well-known carcinogen. A 12-ounce can of diet soda contains 180–200 milligrams (about 4–5 packets) of aspartame, which constitutes about 90 mg of phenylalanine, 72 mg of aspartic acid, and 10–20 mg of methanol.

In the midst of all these mixed reviews about artificial sweeteners and sucrose, are there alternate natural sweeteners that could be safe for our health and much sweeter than synthetic sweeteners?

Yes, there is something new out there that most do not know about: sweet proteins. You might be thinking, proteins are not thought to be sweet. Whenever we hear the word "proteins," we usually relate it to an animal product, such as meat, milk, or cheese. However, sweet proteins are less common in nature. They are produced by a few tropical plants—all come from the rain forests of Africa and Asia. One is commercially available (thaumatin) and approved by the FDA for specific use to modify and enhance flavors (FEMA GRAS Number 3732, "GRAS" meaning "generally recognized as safe"). Others are highlighted for future markets. Table 4-1 describes the source plants, their geographic distribution, and the chemical name of the sweet proteins derived from them.

Table 4-1. Sweet protein, source plants, and their geographic distribution

Tropical plant and geographic distribution	Sweet protein
Dioscoreophyllum cumminsii; Diels, West Africa	Monellin
Pentadiplandra brazzeana; Baillon, West Africa	Brazzein
Capparis masakai; Levl, China	Mabinlin
Thaumatococcus danielli; Benth, West Africa	Thaumatin
Curculingo latifolia; Malaysia	Curculin
Richadella dulcifica; West Africa	Miraculin (sweet taste modifier)
Pentadiplandra brazzeana; Baillon, West Africa	Pentadin

Sweet proteins are very sweet. Most of them are 100 or even 1,000 times sweeter than sucrose—the simplest sugar. These sweet proteins could be especially beneficial to people who are prone to obesity and diabetes and those of us who consume high-calorie, sugar-based drinks and foods. It is worth noting that the obesity rate among children aged 6–11 has increased from 11.3% in 1988-1994 to 17.5% in 2011-2014, and as of 2014, diabetes affects 9.3% of the US population. Sweet proteins have the potential to be used as sweeteners in common foods without leading to the negative metabolic effects that sugar causes.

Sweeter Than Chocolate?

There are seven known sweet proteins (see Table 4-1 and Figure 4-1), many of which are so sweet that only an extremely small amount would taste the same as, say, a piece of chocolate. The main benefit is that they contain a negligible amount of extra calories and no known negative metabolic effects, for the same satisfaction level.

What are these sweet proteins, and why are they sweet? Let's take a look at three of them: brazzein, thaumatin, and monellin (Table 4-2).

Of all the known sweet proteins, brazzein is the most promising, because it tastes like sugar and maintains its structure over a wide range of temperatures and various pH levels. Brazzein comes from a climbing berry plant that grows in West African countries such as Angola, Gabon, Congo, and Nigeria. It is the smallest of the sweet-tasting proteins and is 500–3,000 times sweeter than sucrose (for equal weights). It is not available in stores yet, but will be marketed as a sugar substitute under the name CweetTM if the FDA clears it as safe for consumption. Additionally, a genetically modified corn engineered to produce this protein is under development. This corn would produce a noncaloric, sweet flour for dietetic and diabetic markets.

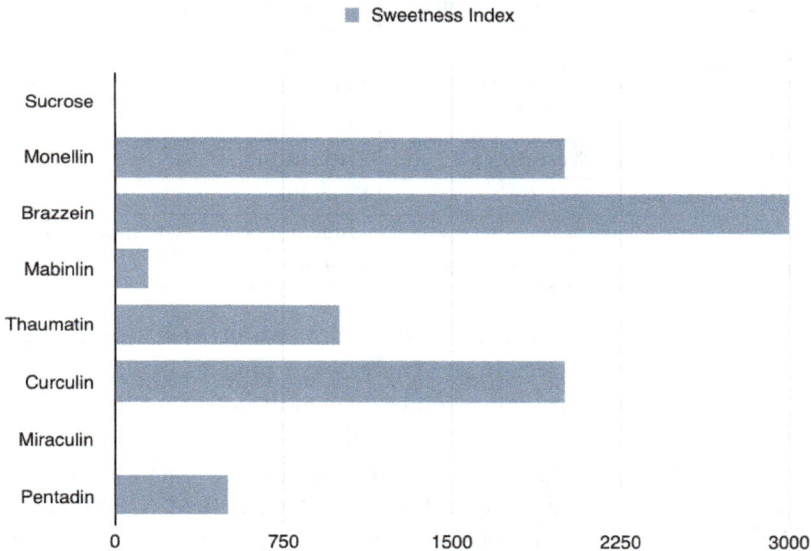

Figure 4-1. Comparison of sweetness levels of the seven known sweet proteins to the sweetness of sucrose (arbitrally set to 1 here, for comparison purposes).

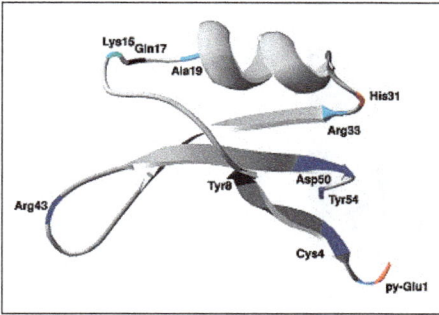

Figure 4-2. Chemical structure of brazzein (also called a ribbon diagram), a monomer protein (54 amino acids) showing the positions of amino acids those are responsible for the sweetness of the protein. Color code: red, enhanced sweetness; light blue, moderately decreased sweetness; dark blue, strongly decreased sweetness; dark gray, low sweetness. (Reproduced with permission from Dr. John L. Markley, University of Wisconsin-Madison)

Thaumatin (Figure 4-3) is another sweet protein that was discovered in 1972 and comes from a tropical plant native to West Africa. This protein is 1,000 times sweeter than sucrose. Thaumatin has extensive disulfide cross-linking, making it particularly stable at high temperatures. This property is especially important if thaumatin is a part of a dish that requires high heat, such as in baking, roasting, grilling, and broiling.

Thaumatin is official as a sweetener in Europe and Japan. In the US, where it is marketed under the name Talin, it is considered safe. The Flavor and Extract Manufacturers Association of the United States (FEMA) gave it a GRAS status (GRAS meaning "generally recognized as safe") for use as a flavor enhancer in chewing gum and breath fresheners and also as a general use applied to all food. In the European Union, it is approved for use in confectionary and chewing gum. In Japan, thaumatin is used as flavor enhancer in most food and beverage industries.

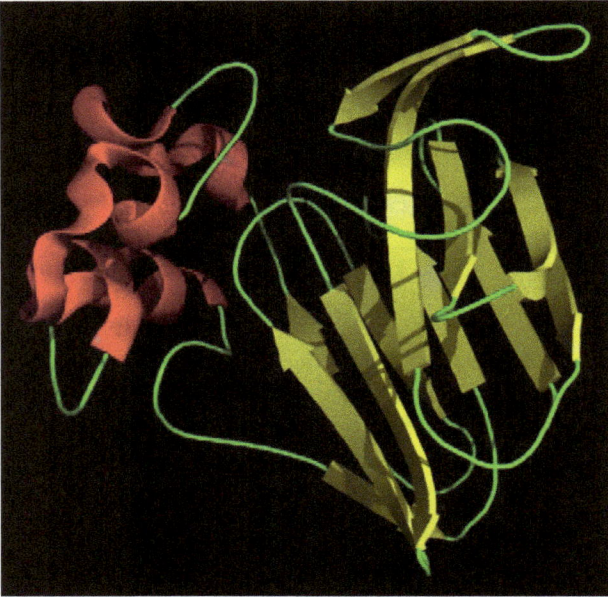

Figure 4-3. Chemical structure of thaumatin monomer protein (207 amino acids). Credit: public domain (https://goo.gl/7UnPRg).

Animal studies have shown that thaumatin had no effect on blood glucose or body weight. However, more research is needed to confirm these results. Thaumatin also shows promise to improve oral health, specifically dental cavities. Natural Health Organics, an Australian (New South Wales) herbal company is marketing toothpaste with thaumatin as a sweet ingredient.

Monellin (Figure 4-4), another sweet-tasting protein, was discovered in 1969 (the first sweet protein to be discovered) in serendipity fruits, which are native to West African countries. Monell Chemical Senses Center in Philadelphia determined that it is a sweet protein and not a carbohydrate. Even though monellin was discovered in 1969, it has not received approval in the food industries of the United States, whereas Japan has approved it as a food additive.

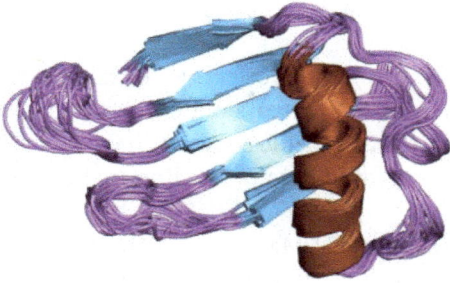

Figure 4-4. Chemical structure of Monellin. A heterodimer consisting of A chain (45 amino acids) and B chain (50 amino acids). Public Domain. Courtesy of Jawahar Swaminathan and MSD staff at the European Bioinformatics Institute. https://commons.wikimedia.org/wiki/File:PDB_1mnl_EBI.jpg and courtesy of http://www.ebi.ac.uk/.

Table 4-2. Sweet proteins, brand names, relative sweetness, and uses

Protein	Brand name	Relative sweetness	Uses
Thaumatin	Talin (US)	1,000X	Approved as a sweetener in Europe and Japan, but in the US, it is considered as safe for use only in chewing gum, breath fresheners, and as a flavoring agent. Not approved as a sweetener in the US or Europe.
Monellin		2,000X	It is approved as a food additive and sweetener in Japan. Used as a table-top sweetener. Large-scale production of monellin has been a challenge due to its heat and pH sensitive properties.
Brazzein	Cweet (awaiting regulatory approval since 2008)	3,000X	ProdiGene (a spin-off of Pioneer hybrid, Inc.), a genetic engineering company in Texas, is developing a sweetened corn flour for diabetic and dietetic market. Natur Research Ingredients, Inc., in California is trying to manufacture and market brazzein as a sweetener for dietetic and diabetic market under the trade name Cweet. The company has not yet received FDA approval.

What Makes Proteins Sweet?

When studying the chemical structures of the seven known sweet proteins, scientists saw that they all have an abundance of beta strands (see Figures 4-2, 4-3, and 4-4). Besides that observation, scientists have not been able to identify a "sweet" part that would be common to all of them. Instead, it seems each protein is sweet for a different reason.

For example, studies of brazzein protein by Assadi-Porter and his team at the University of Wisconsin-Madison have shown that when some amino acids are removed from its chemical structure, it loses its sweetness, which indicates that these amino acids are essential for enhancing sweetness by binding to the sweet tasting receptors in our tongue.

A recent study of monellin by Japanese scientist Masanori Kohmura and his team has shown that monellin's sweet taste is due—at least in part—to an amino acid called aspartic acid (Asp[57]) that is for making an ionic interaction with a taste receptors on the surface of our tongues.

Scientists are just beginning to solve the mysteries of these sweet proteins and how they induce the sweetness on our tongues. All these studies promise to help us understand why we crave sugary substances and why some people want sugar more than others. Customer demand will lead to new types of foods that do not contain as much sugar and that would ultimately help people reduce their consumption of it. Therefore, sweet proteins might play a pivotal role both in food and taste research in the future. And since it has been shown that too much sugar is detrimental to one's health, they might be a viable substitution for sugar in foods. But, you may say, we just learned about the existence of sweet proteins; how nice it will be to taste a brazzein- or monellin- supplemented diet ice cream. Yes, you are right—variety is the spice of life, and innovation to reduce the amount of sugar in food, while maintaining product integrity, is an absolute necessity for good health. You can have your cake and eat it too!

Selected References

- Assadi-Porter, Fariba M., David J. Aceti, and John L. Markley. 2000. "Sweetness Determinant Sites of Brazzein, a Small, Heat-Stable, Sweet-Tasting Protein." *Arch Biochem Biophys.* 376, no. 2: 259–65. *https://goo.gl/QSXoyB.*

- Bahndorf, D, & Udo Kienle. 2004. World Market of Sugar and Sweeteners. International Association for Stevia Research e.V., April 17. *https://goo.gl/q6y5ox.*

- Flatt, J.P. 1970. "Conversion of Carbohydrate to Fat in Adipose Tissue: An Energy-Yielding and, Therefore, Self-limiting Process." J. Lip. Res. 11:131-143. *https://goo.gl/9PBhLA.*

- Khayata, Warid, Ahmad Kamri, and Rasha Alsaleh. 2016. "Thaumatin Is Similar to Water in Blood Glucose Response in Wistar Rats." *Int. J. Acad. Sci. Res.* 4 (2): 36–42. *www.ijasrjournal.org.*

- Kohmura, Masanori, Toshimi Mizukoshi, Noriki Nio, Ei-ichiro Suzuki, and Yasuo Ariyoshi. 2002. Structure-Taste Relationships of the Sweet Protein Monellin. *Pure Appl. Chem.* 74, no. 7: 1235–1242.

- Lindley, Michael. 2012. "Natural High-Potency Sweeteners." In *Sweeteners and Sugar Alternatives in Food Technology, Second Edition. Eds. Dr. Kay . O'Donnell and Dr. Malcolm W. Kearsley. Oxford: Wiley-Blackwell.*

- National Academy of Sciences. *Sweeteners: Issues and Uncertainties.* 1975. Washington, DC. *https://goo.gl/Dedi51.*

- National Center for Health Statistics. *Health, United States, 2004: With Chartbook on Trends in the Health of Americans.* 2004. Hyattsville, MD. Accessed July 8, 2011. *https://www.cdc.gov/nchs/data/hus/2015/059.pdf.*

- National Diabetes Information Clearing House. Accessed July 8, 2011. *www.diabetes.niddk.nih.gov.*

- Pages, Patrice. 2008. "Tasteful Chemistry." *ChemMatters* 26, no. 4: 4–6.

- Vyas, Ankur, Linda Rubenstein, Jennifer Robinson, Rebecca A. Seguin, Mara Z. Vitolins, Rasa Kazlauskaite, James M. Shikany, Karen C. Johnson, Linda Snetselaar, and Robert Wallace. 2014. "Diet Drink Consumption and the Risk of Cardiovascular Events: A Report from the Women's Health Initiative." Journal of General Internal Medicine 30, no. 4: 462–68. doi:10.1007/s11606-014-3098-0.

Dr. Raj Nagarajan, a plant biochemical geneticist from Kansas State University, established his research and teaching occupations with various organizations in US and Canada. Currently, he is a science educator with Milton Hershey School at Hershey, Pa, a philanthropic institution established by Milton & Catherine Hershey that serves students from a disadvantaged background. Dr. Nagarajan can be reached via email.

IndieBio Demo Day 2016

Glen Martin

This spring's IndieBio Demo Day, held as usual at the Folsom Street Foundry, felt distinctly different from its predecessors. The microbrews still flowed freely if not excessively, and the finger foods were elegant and savory. And as with past Demo Days, the place was packed. People were standing cheek by jowl, save for the fact that very few jowls were in evidence among the largely under-35 crowd.

But the ambience had somehow shifted. Earlier Demo Days were charged with a giddy anticipation of a great but unknown future. The air seemed to crackle with static electricity and ozone. Comparisons with Silicon Valley in the early 1980s were widespread, and they felt apt.

But in the latest event, there was a pervading feeling that DIY biotech was maturing at an accelerating rate. The technology remained as surprising and impressive as ever, but the ancillary infrastructure needed to embed those ideas and prototypes into the real world now felt imminent.

That conceit is IndieBio's raison d'être. The San Francisco-based biotech accelerator's mission is to transform diffident and perhaps socially awkward scientists into articulate entrepreneurs who can explain their products and services and attract capital.

"We're essentially running a four-month boot camp, a really rigorous business school for scientists," said Ryan Bethencourt, IndieBio's program director. "Do they love it? No. A lot of them hate it, because we take them way outside of their comfort zone. We make them do cold calls, over and over. We make them do follow-up. We make them close deals. At first they choke. But we keep at them until they succeed. Because it's not enough to come up with a groundbreaking process or product. There has to be a commercial application, and it has to be brought to the world. There have to be customers. And the best people to make those pitches are the scientists, the creative people responsible for the breakthroughs. That way, they not only control the way the technologies are developed and presented, but they're better able to protect their own interests."

Bethencourt said IndieBio is developing a dense network of partners, including corporations (Roche, Bristol-Meyers Squibb, and Unilever) and universities (Stanford, Harvard, Cambridge, and UC Davis.) Bethencourt and company also have discussed possible partnerships with representatives from Singapore, Russia, Canada, South Africa, Taiwan, Chile, and Turkey.

"It's becoming increasingly clear that the classic biotech model is fragmenting," said Bethencourt. "The wide availability of powerful, cheap technology means that huge, centralized facilities are no longer needed for world-changing breakthroughs. And the 'Postdocalypse'—the thousands of extremely intelligent and highly educated bioengineers who find themselves working as lab techs for $45,000 a year due to a dearth of research and academic positions—means that we have the people who can make use of that technology, people who have nothing to lose by going out on their own. Future biotech is quick, cheap, powerful, decentralized, and responsive to changing needs in a changing world. IndieBio is the nexus for these trends."

Venture capitalists are certainly interested, evidenced by their turnout at IndieBio's third Demo Day; at least 400 investors attended the presentations. "I think that we're incredibly fortunate to be participants," said Shahin Farschi, a partner at Lux Capital. "The challenge in technology has always been making the transition to business. IndieBio is providing a tremendous service by helping founders get ready for prime time, and creating the forum that allows investors to get their heads around these new companies. They're creating narratives about what these companies can be, and they're narratives investors can understand. They're not only helping founders create value; they're helping investors recognize value."

Fourteen startups made presentations at the spring 2016 event, representing the medical, food, and cosmetic sectors. All the presentations were well-received, and the following seemed to generate particular enthusiasm.

mFluiDx

Recent pandemics of diseases such as Zika and Ebola highlight a profound deficit: cheap, rugged, accurate, and easily deployable field-based diagnostic devices.

"Right now you have to choose between cartridge-based benchtop diagnostic systems and paper-based field tests," said mFluiDx CEO and cofounder Charlie Yeh. "The first is highly sensitive but very expensive. The second is cheap but sensitivity isn't that good, particularly when it comes to serial conversion; you can't detect infection within a week of exposure."

mFluiDx offers a "valuable middle ground," said Yeh: a DNA diagnostic field test the dimensions of a stick of gum that provides highly accurate results—including for serial conversion—in 30 minutes. The test's self-powered disposable chip costs 1,000 times less than benchtop-based lab diagnostic systems.

"We're working with Gubio Campos (a virologist who contributed to the identification of Zika in Brazil) to show the capabilities of the core technology with Zika and dengue fever," said Yeh. "Ultimately, this will be a platform that will allow quick and accurate field tests for multiple targets."

Qidni Labs

There are 650,000 people in the US suffering from kidney disease; about 100,000 of them are on the waiting list for a kidney transplant. But demand for kidneys far outstrips the supply, meaning that the future for most of these patients is long-term dialysis. That confines each of them to a machine three times a week for four hours per session. Dialysis also is expensive, typically costing more than $80,000 a year per patient. Further, fully 65 percent of dialysis patients can expect to die within five years of starting treatment.

These grim statistics are the driving force behind Qidni Labs' efforts to develop a fully functional and implantable artificial kidney. So far, things look promising. "We've completed two tests with our implantable device in porcine models, and we've been able to establish blood flow and remove solutes from the blood," said founder Morteza Ahmadi. "Those are major derisking steps. We're planning about 50 additional animal tests over the next two years, working closely with clinicians and the scientific community. Once we demonstrate both safety and long-term efficacy, we should be ready to connect to a human patient."

Willow Cup

Alternative dairy products have their good points: little if any saturated fat, less environmental impact because they're plant rather than bovine sourced, and no ancillary animal welfare concerns. Still, the drawbacks are significant. Most pertinently, a widely held opinion that the products aren't all that palatable. Almond and soy milk taste little like real milk, and are less than stellar as cooking ingredients. And tofu "cheese"? Well, the less said, the better.

Willow Cup is aiming to change that, hacking plant-based proteins to create sustainably sourced, cruelty-free alt dairy products with the luxuriant mouthfeel, rich taste, and cooking properties of bona fide milk and cheese.

"We're breaking down the particle size of plant-based proteins and fats and creating matrices that are almost identical to those of cow dairy products," said

Sara Bonham, Willow Cup's CEO and cofounder. "We're obtaining isolates from a number of sources—including chickpeas, potatoes, and cashews—to optimize our formulations. There are a lot of good reasons to opt for a plant-based dairy product, but we don't think a terrible cappuccino has to be the price you pay for that choice."

MycoWorks

Leather is wonderful stuff, but lots of cows and pigs have to die to make all those shoes and jackets, a fact that bothers many people. Synthetic substitutes such as pleather and naugahyde are obviously spurious, and don't feel good against the skin. So what to do?

Turn to the humble fungi. MycoWorks produces "mushroom leather" from mycelia, the fibrous threads that all fungi produce.

"Mycelium leathers are 'skin' in their own right," said MycoWorks CEO and cofounder Sophia Wang. "We induce the mycelia to grow in the matrices we want, obtaining different qualities with different interventions, including light. These mycelia layers are literally alive, and we can obtain different thicknesses and shapes by putting the different layers together and letting them bond. We then put the sheets through a tanning and curing process."

Unlike animal-based leather, mycelia leather is not constrained by size.

"We've made sheets 27 feet square," said Wang. "You could fabricate rectangles, circles—even cowhide shapes."

Mycelia leather could ultimately replace many true leathers and polymer foams, said Wang, who observed her company is in the process of scaling up to a large volume pilot facility.

"We've put our leathers through mechanical and strength tests, and they outperform synthetic, sheep, and deer leathers," Wang said. "Like real leather, they're gas permeable—'breathable.' In our next round of tests, we expect them to compete with bovine leather for strength and toughness. And our inputs, of course, are minimal. We don't have to feed and care for animals. We don't have to kill animals. Our mycelia all are grown on agricultural and wood waste."

Companies like MycoWorks, Qidni Labs, mFluiDx, and Willow Cup point to a sea change in biotechnology, said Bethencourt.

"We're showing that the way to prosper in the future is to embrace change, to be quick and nimble and ultra lean," he said. "The opportunities for both innovative founders and savvy investors are immense. We're just trying to bring them together."

Glen Martin covered science and the environment for the San Francisco Chronicle for 17 years. His work has appeared in more than 50 magazines and online outlets, including Discover, Science Digest, Wired, The Utne Reader, The Huffington Post, National Wildlife, Audubon, Outside, Men's Journal, and Reader's Digest. His latest book, Game Changer: Animal Rights and the Fate of Africa's Wildlife, was published in 2012 by the University of California Press.

Synthego: On the Forefront of Genome Engineering

Sindhu Ravuri

The emergence of CRISPR (clustered regularly interspaced short pal-
indromic repeats) facilitated modification of any singular gene, exposing it
to further scientific analysis and revealing its organismal consequence. As
a result, CRISPR technology not only introduced a higher standard of spe-
cificity to genome engineering, but also shifted the field's paradigm dra-
matically. Today, thanks to CRISPR, scientists can make particular
changes to the DNA of plants, animals, and other organisms—from gene
deletions and knockdowns to insertions and single nucleotide polymor-
phisms—in a more efficient and accessible fashion.
A key component of the CRISPR process involves RNA reagents. Scientists
design the RNA, ensuring each molecule matches a target DNA sequence.
These newly formed RNA molecules then guide Cas9 to the aforemen-
tioned sequence (Cas9 being a nuclease responsible for cleaving the DNA).
*   One of the leaders in synthetic CRISPR reagent development is Syn-*
thego. With Synthego's online tool, scientists can easily design such RNA
reagents for engineering across organisms in various experiment contexts.
We interviewed them to discover more about their origins and technology.

Synthego is a virtual laboratory designed to ultimately provide biology as a service, conducting millions of experiments simultaneously. The technological platform not only creates valuable "parts" for biological research, but also enables engineers to construct better tools. By incorporating customer design with product accessibility through an online platform, Synthego is one of a few biotech companies tapping into an unprecedented market, according to CEO Paul Dabrowski.

Started in 2012 by Paul and Michael Dabrowski—both previously SpaceX engineers—Synthego initially hinged on innovating the tools needed to facilitate a scalable, automated system for biology as a service. Eventually, the company transitioned to its current focus, the reagents market.

"We realized that synthetic biology, biotech, and personalized medicine were going to solve the most important problems in the world this coming century," said Dabrowski. "There's opportunity, yet the tools that researchers have available are really poor. ... Right now, we're focused ... on basically creating kits for CRISPR, because there's a huge need for better tools and products."

Interestingly, this is not the first CRISPR kit on the market. One can purchase a kit from The ODIN (*http://www.the-odin.com/diy-bacterial-gene-engineering-crispr-kit/*) for bacterial cell engineering. The purpose of the kit The ODIN is selling is to teach the basic molecular biology techniques required for CRISPR, so it is an example experiment. What Synthego is providing is a custom kit to engineer any target sequence in a variety of organisms. It's more for the industry and academic lab wanting to conduct experiments more quickly and efficiently. Meanwhile, the ODIN is more for the at-home, DIY user learning new techniques. Although the audiences are different, both kits are making CRISPR technology more accessible.

The Synthego CRISPR/Cas9 EZ RNA Kit is designed to provide results to the customer for a specific gene edit. This kit comes in two separate formats, cr:tracrRNA and sgRNA, each relying on a unique and distinct type of synthetic guide RNA. The kit can be purchased with or without Cas9, and is armed with all the ingredients necessary to "transfect, target and edit."

The synthetic "sgRNA is important because you can really maintain the quality—if you have high quality, input-secure experiments, you get high quality results," Dabrowski said. "With direct synthesis, and the technology that we created to make that happen, you can have very reliable, repeatable results. We're talking 90%, sometimes even 100% efficiency."

This kit will arrive in the mail, along with a protocol card detailing how to actually combine a couple genes together. The customer would then have to use a transfection reagent or perform electroporation to engineer the cell, mammalian or plant. "That's the general [EZ] workflow—so this can basically be done the same day as you receive your kit. We're talking about making CRISPR experimentation available within a couple of days." In addition to these available tools, one can order a Custom RNA kit for varying sgRNA sequence length.

The MVP of these kits, and what Synthego is referring to as CRISPRevolution is the synthetic sgRNA, stabilized sgRNA that is optimum for editing efficiency (CRISPRevolution is a portfolio of synthetic sgRNA, and a hallmark of Synthego).

"With synthetic RNA in particular, we're able to purify to a very high quality, which means that you end up with only one type of molecule in your result—your target molecule."

This, Dabrowski says, ensures Cas9 is not binding to contaminants, therefore decreasing editing efficiency.

What is evident through these products is Synthego's brand focus on synthetics; its online accessibility shortens the time it would usually take to produce IVT (in vitro transcribed) or plasmid-derived guides, for instance. The synthetic guide RNA has the potential to render "pre-sgRNA" techniques useless by cutting down the wait time to receive reagents and also enabling precise editing even with more "challenging cell types" where frequent adjustments are essential. According to their team, there are multiple factors that set Synthego apart from competitors who also provide synthetic RNA reagents. One being the "precision, automation and throughput in the synthesis of the synthetic RNA." The second is increased editing efficiency, resulting from the purity of RNA, with a 90% reduction in off-target editing. And finally, Synthego has reduced the cost by up to five times and shortened the turnaround time to receive reagents by up to four times.

"We wanted to move away from a manual way of doing CRISPR and ... [ultimately] provide world-class research tools available to all researchers in molecular biology and biotechnology, that enable rapid, accelerated discovery," Dabrowski said. "Right now, we are in a situation where the scientists essentially tell us which targets they're interested in modifying, and we create the kits that enable them to do that modification. So they are doing a decent amount of the design work. In the future, you can envision that we actually take on that responsibility."

The capabilities of this product, according to Dabrowski, allow CRISPR to be comprehensively and extensively employed. For example, "if you're trying to modify in another area like personalized medicine, or immunotherapies, you want to modify very hard targets like stem cells. To do so, you need a really high quality, high consistency editing stream. So we're not only making these research tools accessible, but we're trying to make it so that there's even more value to all these downstream applications," said Dabrowski.

Although the audience of the kits is largely on the applications side, Synthego's audience dips into no specific scientific discipline, due to the company's multidisciplinary approach.

"In this field, research is a very interesting term, because it actually spans from academia, all the way through industry, and to the application side. So it's a worldwide scientific community," Dabrowski said when asked about the target audience. "We're seeing various collaborators who are working on immunotherapies to cure cancer; or who are doing research on multiple sclerosis and knocking out the genes associated with that." Today, Synthego's audience expands over 50 countries worldwide. "People are finding us all over the world because we have a unique product and it really makes CRISPR much easier," Dabrowski said.

It is Synthego's novel synthetic approach to CRISPR reagent synthesis, according to Dabrowski, that distinguishes it from its competitors. Dabrowski and his cofounders built a team of engineers who are enabling the development of better tools. The philosophy they have is to provide a way for researchers to be more productive by focusing on providing the tools and reagents necessary. CRISPR is still in its infancy, and the potentials of the technology are unfolding and evolving; but having cost-effective and efficient tools in place will get us to those potentials more quickly. What they are doing is making CRISPR more accessible by bringing down the cost of these tools, reducing the time it takes to receive them, providing everything necessary to edit your gene of interest, and increasing editing efficiencies.

Having such resources will help revolutionize CRISPR even further and can empower more researchers to advance biological research, extending its reach to every echelon of society.

Sindhu Ravuri is a sophomore intending to major in molecular and cellular biology at the University of California, Berkeley. Currently, she is currently one of five national Active Voice journalism Fellows for the Student Press Law Center, combating censorship of young women pursuing STEM fields. Her features on science and activism have earned her a first place for research at Cambridge University's Triple Helix conference, a national YoungArts Award, and a literary scholarship at UC Berkeley.

Genes in Space

Julian Rubinfien

For a high school student interested in biology, there are more educational opportunities available outside the classroom than one could possibly experience in four years of American secondary education. There are summer internships at universities, essay contests, scientific scholarships for college, and—for those with access to professional laboratories and mentorship—science fairs. Biology competitions are generally more alike than they are different; most involve a lengthy application and judging process that culminates in the presentation of cash prizes to winners. I was surprised, then, when I read that Genes in Space (*https://www.genesinspace.org*), a two-year-old competition, promised its winner the opportunity to launch his or her biology experiments to the International Space Station (ISS), a collaboration between 15 nations that orbits the Earth at 17,000 miles per hour. The Genes in Space competition is a partnership between miniPCR (*http://www.minipcr.com*), Math for America (*http://www.mathforamerica.org*), the Center for Advancement of Science in Space (*http://www.iss-casis.org*), New England Biolabs (*http://www.neb.com*), and The Boeing Company (*http://www.boeing.com*).

The idea of outer space is irresistibly romantic. There is no place more extreme, more foreign, or more beckoning. In fewer than 60 years, a total of 536 men and women have called space a (temporary) residence. Since then, we have been, as a species, reaching toward the future, working constantly to push the limits of habitability. From Lego bricks (brought to space by Japanese astronaut Satoshi Furukawa) to Marshmallow Fluff (stocked in the ISS kitchen by American astronaut Sunita Williams), we've slowly turned outer space into a home for humans. More recently, space has become a home for laboratory science. The expansion of biotech into space has become tremendously important as the world prepares itself for mankind's next giant leap, the first manned mission to Mars, and confronts the undeniable fact that we are not yet ready—technologically, financially, or physiologically—for long-term space travel. Enter the ISS National Laboratory and the Genes in Space competition.

There are numerous health problems associated with life in space. Loss of muscle mass and bone density (see here (*http://europepmc.org/abstract/med/8747608*), here (*http://www.ismni.org/jmni/pdf/2/leblanc.pdf*), and here (*http://bit.ly/2nVSdbe*)) in astronauts receives a great deal of press, but the rigors of spaceflight more generally induce a number of physiological, genetic, epigenetic, transcriptomic, and metabolic changes in the human body, not to mention the psychological stresses of living in such a hostile environment. Interestingly enough, many physical symptoms experienced by astronauts mimic symptoms of aging among the elderly on Earth: increased risk of cardiovascular disease and various cancers, for example. These trends have generally been observed in astronauts living aboard the ISS.

In order to research changes in human health during spaceflight, NASA and its partner organizations have established various laboratories on the ISS that are capable of analyzing astronaut health on site. As researchers delve deeper into the molecular causes of spaceflight-associated illnesses, these laboratories are evolving to facilitate molecular biology research. In recent years, if NASA scientists wished, for example, to analyze changes in the expression of a set of genes over the course of a space mission, astronaut blood samples would have to be taken during the flight, frozen, and then shipped back down to Earth for analysis; the ISS laboratories are not currently equipped for any kind of advanced biomolecular research. For more complex, long-term studies, multiple flights to and from the ISS would be required. This obviously presents a financial and technical bottleneck for advanced research into astronaut health, and it is a problem that the Genes in Space competition is helping to solve.

Over the evolution of laboratory science in space—various Russian space stations, one Chinese space station, NASA's Skylab, and now the ISS—the demand for extraterrestrial analysis of astronaut health has spawned a kind of cottage industry based on research conducted by astronauts on astronauts in low Earth orbit. In some ways, the challenges of bringing science out of planet-based academic laboratories and into space aren't unlike the challenges faced by DIYbio enthusiasts seeking to bring science out of those insular enclaves of research that include universities and biotech companies. Functionality, portability, accessibility, and cost-efficiency are the most important qualities of laboratory equipment that's launched to the ISS. Many standard pieces of lab equipment simply won't work in microgravity. And still, America's own space lab, the ISS National Laboratory, is being upgraded every day. Recently, for example, the National Lab was equipped to care for rodent astronauts; researchers have been attempting to develop the use of mice as a model system to study human health during spaceflight.

And so, when I applied to the national Genes in Space competition (along with hundreds of other high school students), I was invited to propose a set of molecular biology experiments that could be performed in the ISS National Lab and might help answer some of the most pressing questions about human health in space. The invitation came, of course, with a unique set of constraints. Proposed experiments had to be original and scientifically relevant, but they also had to be feasible; they could not employ recent, popular scientific techniques like nucleic acid sequencing or flow cytometry, because the National Lab simply doesn't have the required equipment. When I was developing my proposal, I tried to think of what current space scientists were interested in with regard to astronaut health and what experiments I could propose that, if performed, would expedite and simplify their research.

I chose to center my proposal around the study of human telomeres—vital protective caps on the tips of chromosomes—in space. Abnormal telomere lengths, whether too long or too short, have been associated with a remarkable number of human diseases in various tissues. Generally, if telomeres become critically short, cells will enter a state of senescence in which cell division ceases. As human bodies age, senescent cells naturally build up in various tissues. The connections between physiological stress and aberrant regulation of telomere length have been well documented. It is hypothesized by scientists that the stresses astronauts experience during spaceflight, such as poor quality of sleep, inadequate nutrition, microgravity, and cosmic radiation, may lead to changes in telomere dynamics during space missions, and in turn, to tissue-specific senescence and symptoms of disease. This hypothesis is currently being tested by the ongoing NASA Twins study (*https://www.nasa.gov/twins-study*), a comprehensive analysis of the changes that occurred in astronaut Scott Kelly's body over the course of nearly a year in space.

Since the ISS National Lab's resources are so sparse, the need for tools for research is almost more urgent than the research itself. When developing my proposal, I was concerned primarily with the measurement of changes in telomere dynamics. I proposed the in-space validation of an assay which uses traditional polymerase chain reaction (PCR) technology to measure changes in telomere lengths. The assay I want to use was developed to measure telomere lengths (*http://bit.ly/2pKc6P6*) on Earth by amplifying and visualizing them (Figure 7-1), but—with specific modifications—is uniquely suited to use on the ISS.

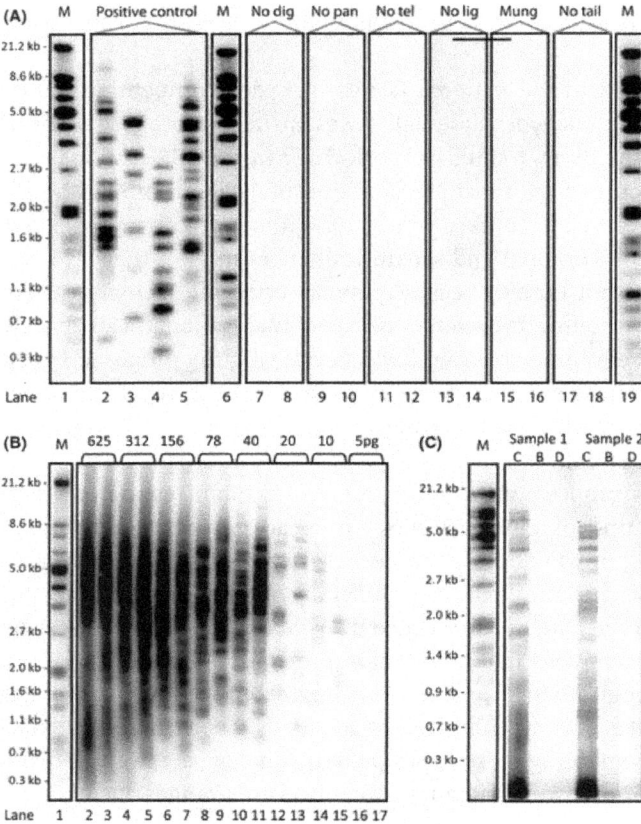

Figure 7-1. The single telomere length analysis, STELA, is a ligation-PCR-based method to meas-ure telomere length (from Bendix et al., 2010.)[1]

All the same, part of the proposal was imagining how others could build upon my experiments after their completion. With the generous assistance of my mentor, Deniz Atabay (*http://bit.ly/2pK3VlV*), an MIT PhD student in neurobiology, I designed a hypothetical study of telomere dynamics in a human organoid model system, cultured human cells that are capable of forming primitive organ structures, aboard the ISS. The design made use of specific organoids—myocardial organoids, for example, mimic the form of the heart—that could be cultured on the ISS over long stretches of time. Periodically, DNA could be sampled from the organoid cultures, and telomere lengths could be assessed with the appropriate

1 Bendix, Laila, Peer Bendix Horn, Uffe Birk Jensen, Ivica Rubelj, and Steen Kolvraa. "The Load of Short Telomeres, Estimated by a New Method, Universal STELA, Correlates with Number of Senescent Cells." Aging Cell *9.3 (2010): 383-97.* Web.

assay. Such an investigation would allow researchers to answer questions about changes in the regulation of telomere length in different human tissues and organ systems in real time, on site.

In June 2016, I traveled to San Diego to attend the ISS Research and Development conference (*http://www.issconference.org*), my first scientific conference. While there, I presented my proposal to a group of judges and was chosen as the winner of the Genes in Space competition. My experiments, which I've been preparing since I returned home from the conference, are scheduled to be launched to the ISS in late March, 2017. When they arrive, they will be unpacked and executed by an astronaut.

For me and, I imagine, many other high school students, Genes in Space is unique in that its primary aim is not to reward work already done by the student, but to provide the student an opportunity to expand the scope and elevate the significance of his or her work by sending it to the ISS. It's an opportunity that, most of the time, is only available to late-career government scientists.

The value of this opportunity is compounded, counterintuitively, by the state of laboratory research on the ISS. Despite great demand for scientific data obtained aboard the space station, the difficulties of performing experiments while orbiting the Earth mean that outer space molecular biology is still in its very early stages. As a result, a high school student, as unreal as this sounded to me at first, can contribute to mainstream science in legitimate and significant ways. The first polymerase chain reaction, the assay on which all molecular biology is based, performed in space was designed by Anna-Sophia Boguraev (*https://youtu.be/hwok8eYXrDk*), the 2015 Genes in Space winner (*http://bit.ly/2o3FQdu*). My experiments will test the capability of two DNA polymerases to amplify highly repetitive nucleotide sequences in orbit. They will also involve the first colorimetric loop-mediated isothermal amplification of DNA (LAMP) ever performed in space (*http://bit.ly/2on57et*). Genes in Space allowed me to make the jump from my personal science education to the real progression of human knowledge and capability. The novelty of this cannot be overstated.

Julian Rubinfien is a junior at Stuyvesant High School in New York City. He has been studying molecular biology intensively over the last few years and has a great interest in biomedical research. He is the winner of the 2016 Genes in Space competition (http://bit.ly/2pghEEL), in which he proposed a novel method and model organism for studying telomere dynamics aboard the International Space Station, where his experiments will fly in March 2017. He is currently an intern at Memorial Sloan Kettering Cancer Center. Before that, Julian interned at Regeneron Pharmaceuticals. He is also studying advanced Mandarin and enjoys sailing on the Hudson river with his school's racing team.

Are Worms the Future of Drug Development?

NEMAMETRIX'S SCREENCHIP COLLECTS LIVE-ANIMAL, HIGH-THROUGHPUT C. ELEGANS ELECROPHARYNGEOGRAMS (EPGS), WHICH ARE AN INDICATION OF OVERALL ORGANISM HEALTH FOR DRUG DEVELOPMENT.

Courtney Webster, MM DD, 2017

At first glance, studying *C. elegans* feeding seems a bit esoteric. I certainly didn't know that their swallowing frequency is an indicator of age-related health (*http://bit.ly/2prtmfp*), or that monitoring this behavior is used to study neurophysiology and toxicology (Leung, 2008 (*https://www.ncbi.nlm.nih.gov/pubmed/18566021*)) for drug development. Shawn Lockery, the CTO and cofounder of NemaMetrix, compares an electropharyngeogram (EPG)—the measurement of pharyngeal pumping as electric activity—to an EKG. "[You could think of an] EKG as just the heart pushing blood around," he said. "But you can figure out age and health and strength, all the things that are directly correlated to one of [our] most integral organs."

Turns out, the EPG to EKG analogy is not just a literary turn of phrase. The pharyngeal pumping mechanism in *C. elegans* is controlled by a set of genes that determine cardiac health in humans (*http://bit.ly/2oG8LCS*). *C. elegans* possess homologues for up to 80% of human genes (Kaletta, 2006 (*https://www.ncbi.nlm.nih.gov/pubmed/16672925*)) and, as a result, are widely used as a model organism.

NemaMetrix believes that using *C. elegans* as an inexpensive, live-animal model can revolutionize (and democratize) the drug development process. But common methods to analyze pumping rates were archaic and cumbersome. Most often, researchers would just record a video of a worm placed on an agar plate to feed, and then by replaying in slow motion, count its swallows using a handheld counter. EPGs, pioneered by Leon Avery in 1995 (*http://bit.ly/2nVEpO2*), record the electric activity of pharyngeal pumping. This is a more precise way to count swallow frequency and reports the action potentials of the muscles and the synaptic potentials of the neurons controlling the muscles (Figure 8-1).

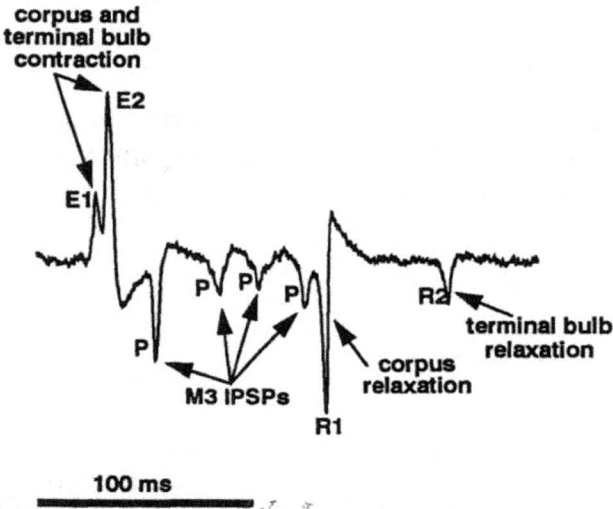

Figure 8-1. A typical normal EPG, in Avery's seminal 1995 publication (https:// www.ncbi.nlm.nih.gov/pubmed/8531728?dopt=Abstract)

But when the ScreenChip technology was first envisioned in 2008, state-of-the-art EPG recordings still relied on manual techniques and (typically) beheaded worms (*http://bit.ly/2prl33m*), making high-throughput and/or live-worm analysis impossible.

So Lockery, a University of Oregon professor studying *C. elegans* and electrophysiology (as well as a self-professed "maker"), imagined a device that could continuously record EGPs from a live worm. This would pave the way for a high-throughput way to measure EPGs.

The key challenge, according to Lockery, was building the right type of channel. The device would need to hold a worm securely in place between two electrodes for the EPG recording, but still allow a flow path so the worm could be exposed to solutions containing any compound of choice (Figure 8-2). At the time, Lockery was about to start a sabbatical with none other than George Whitesides, the father of microfluidics. Lockery and Whitesides lab member Elizabeth Hulme had proof of concept in just a few months. Shortly thereafter, Lockery and University of Oregon biology professor Janis Weeks cofounded NemaMetrix to commercialize the technology, bringing Matt Beaudet on board as CEO.

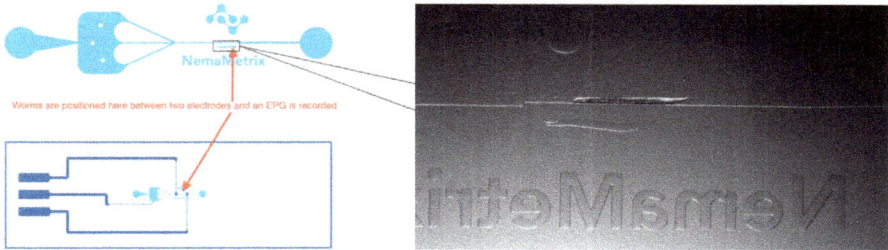

Figure 8-2. The ScreenChip (http://nemametrix.com/our-tech) system uses vacuum to position C. elegans in a narrow channel (https://youtu.be/opCmPmHocPQ) between two electrodes

ScreenChip is designed to be used by any level of researcher and only requires a vacuum, a microscope, and a computer. Adult worms can be bulk-loaded in solutions containing any test compound, and then serially captured for the EPG recording. The system can be used for high-throughput analysis (up to 50 worms per hour) or to record EPGs on a single, live worm for as long as needed. The recorded EPGs are analyzed using open source NemAcquire software. (In fact, if you're having trouble, the website offers to analyze your data for free!)

Though Lockery was interested in commercializing the device from the start, the ScreenChip system was initially typecast as just a handy automation tool for *C. elegans* researchers. Matt Beaudet thought the same thing at first glance in 2012. But then, he said, he began to "mentally overlay [the system] onto one of the big problems I'd been working on for almost a decade: How do you bridge the gap between what a cell biologist and what a mouse researcher can do?" As you progress in the typical drug development pipeline from in vitro experiments to animal models, the cost (*http://bit.ly/2oESw7z*), regulatory oversight, and sheer number of experimental variables make for high stakes. Using ScreenChip, researchers can do inexpensive, quantitative live-animal work that enables them to identify leads quickly and have the rest fail fast before turning to expensive, higher-order animal models.

ScreenChip abstracts the complexity of a live model organism into a database of inputs and outputs, like *in silico* drug development. In Lockery's view, "an animal is just a biological circuit." And the motivation behind the plug-and-play design and open-source software? The founders want to demystify electrophysiology and its analysis. As Beaudet said, "This was a technology that's the most powerful when others are innovating on top of it."

NemaMetrix's sights are set on pharma but are currently focused on the academic research market. Its customers are using the tool for all sorts of research, from models of cardiac arrhythmia and Alzheimer's disease to pharmacology and toxicology. Cofounder Janis Weeks has shown that *C. elegans* exhibits a rapid and concentration-dependent decrease in pharyngeal pumping (*http://nemame trix.com/epg-recordings-new-tool-toxicology-c-elegans/*) after exposure to the heavy metal Cu^{2+} (commonly generated when copper piping corrodes). Lockery has shown that nematodes get hyperphagia (better known as the munchies) after exposure to the cannabinoid anandamide, resulting in well-fed worms doubling their pumping frequency (*http://nemametrix.com/endogenous-cannabinoid-anandamide-causes-hyperphagia-c-elegans*) (compared to the control). NemaMetrix also has nonacademic consumers using ScreenChip to evaluate the health effects of eco-friendly detergents and public school water sources.

Though the core platform has wide applications as is, NemaMetrix wants to turn ScreenChip into a broader platform. Pharyngeal pumping is just one of many relevant *C. elegans* phenotypes for drug development. The company wants to build a system that can measure movement (another important indicator of animal health), as well as fluorescent phenotypes by integrating the system with a small fluorescent microscope. According to Lockery, "Modern medicine didn't stop with EKGs—and we don't plan to stop there either."

Courtney Webster (@automorphyc (https://twitter.com/automorphyc)) is a freelance writer with professional experience in laboratory automation, automated data analysis, and the application of mobile technology to clinical research.

Citizen Health Innovators: Exploring Stories of Modern Health

Eleonore Pauwels and Todd Kuiken

From the Mylan EpiPen pricing scandal, to the whistleblower story that crashed the blood-testing startup Theranos, among many Americans, there is a growing public distrust in governance over the biomedical enterprise and there are questions being raised about who gets access to cutting-edge sophisticated drugs and therapies.

At the same time, there's a parallel story brewing about citizens who decide not to wait to shape their own medical future. One of them is Tal Golesworthy (*http://bit.ly/2paJmCX*), a bright and resolved engineer who, suffering from a genetic disease that damages his heart, designed a surgical device that would save him and other patients from a more risky procedure. Dana Lewis (*https://www.wilsoncenter.org/article/open-artificial-pancreas-openaps-o*), a digital communication specialist suffering from Type 1 diabetes, created an artificial pancreas based on an algorithm that calculates the need for insulin based on a patient's blood sugar levels. And to find a cure for their daughters suffering of the rare Batten disease, a couple (*https://experiment.com/projects/finding-a-cure-for-batten-disease*) raised millions on a crowdfunding platform to hire their own research team. While these individuals and other communities are reshaping their involvement in health research and practice, they are raising new ethical, safety, and governance issues for policymakers, practitioners, and patients.

This participatory turn has no official name. Some say "patient-led" or even "patient-powered" research, others "DIY health." We call them *citizen health innovators* and have begun mapping their emergence and exploring their stories, as well as the ethical and regulatory landscape that surrounds them, with funding from the Robert Wood Johnson Foundation (*http://www.rwjf.org/*) (Map (*https://*

chipmap.wilsoncenter.org/) and Website (*https://www.wilsoncenter.org/ program/citizen-health-innovators-project*)). But how did we get there? What enabled this new societal phenomenon to arise? We identified the convergence of three factors that contributed to a form of democratization in health research and practice: vanishing barriers to entry, the rise of and access to personal genomic data, and the emergence of crowdfunding platforms.

First, the barriers to entry to an array of genetic and biotech techniques have decreased to a considerable extent through PCR machines, gene-editing test kits, and portable genetic sequencers. There is also now the possibility to sequence a genome for about $1,000. Second, biomedical research is increasingly relying on personal genomic data to tailor diagnostics and therapies to groups of patients, creating the incentives for individuals to resort to personal genomics and learn about their own genetic blueprint. The third and possibly most important factor which contributes to this participatory turn is the access to financial backing that citizens recently gained through crowdfunding platforms. After raising about $2,642,000 on *experiment.com* (*https://experiment.com/projects/finding-a-cure-for-batten-disease*), the parents of Charlotte and Gwenyth Gray decided to hire their own research team to accelerate research in three promising treatment options for Batten disease: gene therapy, cellular therapy, and small molecular therapy.

While the convergence of these factors is not necessarily a silver bullet to a cure, it does enable us to imagine one. Which begs the questions, what if it works? And what should the role of government be in these new endeavors? After all, some of these are health conditions and diseases that the traditional research communities have largely ignored or treatments that people cannot afford.

Several governance issues lurk in the background. Compared to standard National Institute of Health (NIH) grants, which can take up to a year to get funded, crowdfunded research can begin in as little as 30 days from when a project is launched on a site. There is currently no official safety and ethical oversight, or a traditional peer-reviewed system that accompanies these proposals, raising complex questions for crowdfunding platforms to tackle. Who's liable when it comes to delivering on the results promised in the funding pitches? Is there a responsibility for the crowdfunding platform to properly vet projects, similar to the NIH peer-reviewed process? Or are they simply a conduit to pass money through with no responsibility, similar to Western Union or bitcoin?

What about the quality of data coming from patient-powered health research? How will traditional academic journals and government agencies assess the data derived from crowdfunded studies that may not have applied NIH rules for health research? If journals and agencies reject such data, does it even matter if the protocols established to produce the treatments and medical devices are accessible to other ends users? Facing regulatory uncertainty, patient innovators might not overcome this "chill factor," a phenomenon described by DIY inventors as the fear to confront regulators by sharing the recipe for a new invention.

The press might cover the few memorable cases of patients who self-experimented (*https://www.technologyreview.com/s/603217/one-mans-quest-to-hack-his-own-genes/*) with unregulated gene therapy treatments. But those are not common practice. As shown on this map (*https://chipmap.wilsoncenter.org/*), patient innovators address crucial user-centered issues with their designs often vetted by peers and doctors who have become collaborators in their shared innovation journey. Nonetheless, we argue that it is important to think creatively about how to help patient innovators share their data, evidences, tacit knowledge, value trade-offs, and ethical concerns in ongoing conversations with regulators and society at large.

We, as a society, are at a tipping point. We could build a new innovation ecosystem that ensures safe and responsible citizens' participation in health research, or we could drive these emerging communities of innovators at the margins, underground or out of existence. What can patients teach us about user-centered research and design? How can regulators help them embed responsible governance mechanisms into their endeavor? How, in turn, can this culture of responsibility confer legitimacy to patient-powered health research?

The goal of the *Citizen Health Innovators Project* is to develop engagement channels with innovators, patients, ethicists, and regulators to design adaptive oversight tools that will foster a culture of empowerment and responsibility. We envision building an open and distributed health innovation ecosystem that empowers patients through tailored inventions and is seconded by adaptive regulatory institutions. This effort to provide patient-led research with more legitimacy is a collective endeavor that needs new practices. Will you join us (*https://www.wilsoncenter.org/program/citizen-health-innovators-project*)?

NOTE *Support for the Citizen Health Innovators Project is provided by a grant from the Robert Wood Johnson Foundation. The views expressed here do not necessarily reflect the views of the foundation.*

Dr. Todd Kuiken *is a Senior Research Scholar with the Genetic Engineering and Society Center at NC State University where he explores the scientific and technological frontier, stimulating discovery and bringing new tools to bear on public policy challenges that emerge as science* advances. He has numerous projects evaluating and designing new research and governance strategies to proactively address the biosafety, biosecurity and environmental opportunities/risks associated with emerging genetic technologies. In September 2016, he along with Eleonore Pauwels, received a grant from the Robert Wood Johnson Foundation to enable the fast-growing ecosystem of "DIY" health innovators to develop a culture of responsibility that reflects its pluralistic and open-source ethos.

Eleonore Pauwels is an international science policy expert, who specializes in the governance of emerging technologies, including genomics and genome-editing, participatory health design, and citizen science. At the Wilson Center, she is the Director of Biology Collectives (https://www.wilsoncenter.org/program/biology-collectives-project), within the Science and Technology Innovation Program. With funding from the Robert Wood Johnson Foundation, and in collaboration with Todd Kuiken at NC-State, Eleonore directs the Citizen Health Innovators Project. Her research focuses on developing governance mechanisms for the fast-growing ecosystem of health innovators, built around maker spaces and community bio labs, to support responsible innovation in distributed networks. She is particularly interested in the perils and promises of personal genomics, and how to harness this trove of data and techniques to truly, ethically empower citizens in different societal contexts and cultures.

Digital Detection: Large-Scale Ambition Built on Micro-Scale Fluidics

MIROCULUS DEMOCRATIZES EARLY CANCER DETECTION WITH AN OPEN RESEARCH DATABASE AND A DIGITAL MICROFLUIDIC PLATFORM

Courtney Webster, MM DD, 2017

Early detection can completely change the trajectory of a cancer diagnosis. We all know by now that the earlier cancer is found, the better. Jorge Soto, the CTO of Miroculus, said in his 2014 TED talk (*http://bit.ly/2oGfAEC*), "Catching cancer early is the closest thing we have to a silver bullet cure against it." When Soto formed a highly interdisciplinary team at Singularity University, they founded their company around this goal.

As cancer develops, it leaves a breadcrumb trail of mutation or disruption. If you are looking for a cancer diagnostic, there are a plethora of options to choose from. You can look at protein expression, DNA mutations, or different types of RNA (noncoding RNA, cRNA, etc.). As it develops, this cancer trail leaks into the bloodstream or urine, where it is easier to sample. When searching for the right diagnostic, you are typically looking for three attributes: Is it easy to get? Is it easy to detect? Is it reliable? Many researchers and companies, including Miroculus, believe that extracellular microRNAs (miRNAs) fit the bill. miRNAs are class of noncoding RNAs that regulate various cellular processes (like differentiation and replication). They make for an attractive diagnostic because they are small, stable, and present in blood and plasma.

The promise of miRNAs for detection is controversial, however. While the best biomarker would be present (or absent) in a disease state compared to normal, miRNAs are instead diagnostic in their concentration (*https://www.ncbi.nlm.nih.gov/pmc/articles/PMC3199035/*). For example, a particular miRNA may be below a certain level normally, but exceed that threshold if the person has a certain type of cancer. To be proven reliable, all other factors that could lead to a significant change in concentration *without* the disease present need to be eliminated (for example, diet, exercise, or other medications). Also, it's been hard to compare miRNA concentrations using different platforms (*https:// www.ncbi.nlm.nih.gov/pmc/articles/PMC3199035/*), or even the same platform with different vendors' products. It's difficult (*http://zon.trilinkbiotech.com/ 2015/11/17/small-rna-big-science/*) to determine exactly how many miRNAs have been approved to date (*http://bit.ly/20G9a8m*) for cancer diagnostics, but to put it simply, few to none

Still, this technique holds incredible promise. With over 2,000 miRNAs in the human genome, it could be possible to develop an miRNA "fingerprint" for any number of diseases (not just cancer). So Miroculus built Loom (*https:// loom.miroculus.com/*), an miRNA data engine, so that its technology can evolve alongside the highly dynamic miRNA research field. Loom is an interactive site (updated monthly) that summarizes literature by miRNA, gene, or condition (Figure 10-1). They've chosen to make this tool open-access for the community, and it plays a key role for the company as well. Instead of funneling resources into a secret discovery effort, Miroculus can use Loom to identify key players in the field for collaboration in addition to its own R&D. Since cancer is one of the most highly funded areas of research, there are plenty of publicly available findings to draw upon. It's important to them that this tool gives back to the community. In their words, more open-access data pushes the science forward more quickly, so everyone benefits.

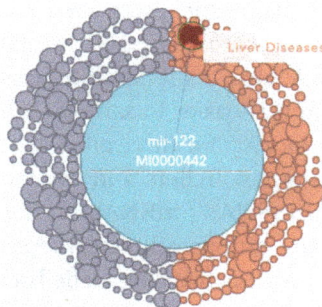

Figure 10-1. Loom.bio's analysis of literature on the microRNA mir-122 (http://bit.ly/20Gbc8M)

After Loom identifies key relationships between miRNAs and a particular disease, Miroculus profiles the miRNA and uses predictive modeling to narrow down the number of miRNAs associated with the disease. Then, this research is validated through clinical studies. Miroculus has just completed a multicenter clinical study on gastric cancer (the third deadliest cancer in the world), and are about complete a second one. The company plans on publishing its findings and submitting the data to the FDA (and other regulatory agencies) for evaluation in the near future.

Loom helps Miroculus identify the "what": which miRNA shows promise as a diagnostic. They also developed new technology to address the "how"—how the miRNA detection is performed. In contrast to other companies, Miroculus's vision is centered around democratization. Standard miRNA detection methods require expensive and specialized equipment, using analytic conditions that require precise temperature control and hot/cold cycling. The stricter the conditions, the more difficult it is for someone to achieve without proper training and proper equipment. Ultimately, Miroculus wants its technology placed in every hospital lab. So it's designed its technology first and foremost to be inexpensive and simple to use.

In Miroculus's first iteration (*http://bit.ly/2oGfAEC*) of their detection platform, it eliminated the precise temperature control required by traditional methods and used a technique that works under isothermal (constant temperature) conditions. Instead of requiring a microscope or expensive plate reader, the output signal is visible to the naked eye with analysis that can be done using a smartphone. As ubiquitous as smartphones are these days, would you still consider them an "everyday" tool? Perhaps in many countries, but certainly not in every hospital lab or clinic globally. So in the second iteration (*https://github.com/miro culus/Miriam*), Miroculus eliminated the smartphone to make the system more robust and even more accessible.

But it's the third iteration (which will be released later this year) that I think is really interesting. Even with cheap equipment and simple conditions, miRNA analysis still requires skill on the part of the technician. For those of us used to working in a lab, pipetting feels like you could do it in your sleep. But that minimizes the importance of skilled hands in producing reliable, reproducible data. Perhaps, though it would be difficult, you could design a method with limited sample preparation that a nonexpert could execute. But for gastric cancer, Miroculus detects miRNA from plasma that has to be processed before the detection step. Instead of redesigning the methodology, Miroculus engineered a plate that manages some of the pipetting for you with digital microfluidics (DMF).

Soto explains digital microfluidics like this: "If you have a board that's printed with electrodes, you can apply a voltage through each electrode that creates a force that makes the droplets on top of the board move. That's how you can move liquids in an automated fashion (*https://youtu.be/d7G_pE8KAN8*)." When I asked him how he selected that method compared to the other microfluidic techniques out there, he said there was really no choice. "This was really the only technique that would accomplish the task in a flexible platform." Using DMF, you front-load all the complexity into the plate. It may seem like this sophistication would result in really expensive starting materials (and heck, isn't the point of democratization to be inexpensive?) But compared to other microfluidic techniques (*http://bit.ly/2pJUmn2*), you can fabricate a general electrode design using simple photolithography to use with DMF. This makes it easy to reconfigure one design for multiple applications. Soto compared it to software: "It gives you the flexibility to adapt as a scientist and makes it easy to make iterations."

While the technology has evolved a lot in the last three years, Soto said the core goal—to democratize molecular testing—has not. We'll be hearing a lot from Miroculus in the next year, and I'll certainly be watching to see what the next iteration achieves.

Courtney Webster (@automorphyc (https://twitter.com/automorphyc)) is a freelance writer with professional experience in laboratory automation, automated data analysis, and the application of mobile technology to clinical research.

IndieBio Demo Day 2017

Glen Martin

At IndieBio's fourth Demo Day, it was clear that a lot had changed. For one thing, the San Francisco biotech accelerator isn't a mere upstart anymore. It has moved its semiannual event, which showcases its partner startups to potential investors, from The Foundry, a hip events center in San Francisco's SOMA district, to the rococo interior of the Herbst Theatre, which can accommodate far more people.

"We just needed a bigger venue," says Ryan Bethencourt, IndieBio's cofounder and program director. "The Foundry only holds 400 or 500 people. The numbers have been increasing with each Demo Day since our first one two years ago, and it was getting to the point that people were feeling like sardines. The Herbst can hold close to a thousand, so we're hoping that will suffice for a while."

That may be overly optimistic. Though the Herbst was sufficiently capacious to accommodate the attendees at the February 19 Demo Day, it still somehow felt like a stopgap. The atmosphere was charged as company representatives presented their products and services on stage; it almost felt that the theater was crackling with ozone. Indeed, the event seemed to point to something larger: the acknowledgement, perhaps, that Next Generation biotech, the smaller, nimbler biotech, the biotech forged in DIY labs and bootstrap accelerators, the biotech that was truly representative of the 21st century, had arrived.

Moreover, observes Bethencourt, the growing interest in IndieBio and its representative ethic is reflected in the most basic and telling metric for capitalist societies: investment dollars.

"When we first started, it was really hard to help our companies raise capital," says Bethencourt. "It was mainly coming from angel investors. Now, two years later, they're raising money from VCs, and the checks are a lot bigger. Further, it's not just local anymore; it's no longer just Silicon Valley investors. We're attracting global investment, and it's been transforming."

Certainly, the Herbst Theatre Demo Day unveiled some impressive products and technologies. Some seemed to promise social or economic transformation. Others were just extremely cool. All drew the deep interest of the scores of investors in attendance.

"Several of the companies in this year's class had significant funding, pre-sales, or both by the time we hit Demo Day," says Bethencourt.

Drawing the strongest interest was Catalog, a company that is building "infinite data archives" in DNA. Presenting for the company, CEO Hyunjun Park noted that data storage is accelerating at such a rate that by 2020, about 100 zettabytes will be archived—or 10 times more bytes than there are stars in the observable universe. Current hard drive technology could well prove inadequate for such demands, Park observes. But with DNA as the storage medium, information densities can be achieved that are 10 million times greater than those afforded by magnetic tape.

"And unlike current technologies, DNA will retain data in an uncorrupted state for thousands of years," said Park.

If information is encoded on DNA's base pairs, then the molecule must be synthesized to meet those specific data requirements—a prohibitively expensive process, Park acknowledged.

"If you had to synthesize enough DNA to archive 1,000 terabytes of data, it'd cost about $240 trillion," Park said. "So we get around that by storing through the absence or presence of DNA molecules. We use prefabricated DNA—no synthesis is required."

Catalog drew $5 million in seed funding by Demo Day, a figure that was not only an IndieBio record, but an amount that will allow Catalog to archive a gigabyte a day by the end of 2017 and a terabyte a day by the middle of 2018. In short, said Park, Catalog's work confirms that DNA is the best extant means for archiving data: "It's the medium perfected over four billion years to store nature's most precious information."

Bethencourt characterized Catalog's initial funding as "somewhat anomalous" for IndieBio, but cited it as evidence that Next Gen biotech "is at the start of its F-15 climb." Virtually all of the 13 companies comprising the accelerator's latest class received investor interest, he observes. Along with Catalog, that included Animal Microbiome Analytics, which is marketing a microbiome-testing service to identify pet maladies; BioInspira, a startup that has developed a virus-based sensor network to remotely detect airborne chemicals in real time; Scaled Biolabs, which has developed microfluidic chips that can support thousands of parallel cell experiments; and Venomyx, a firm that has created the first toxin-specific snake venom.

"The experience of Ravata Solutions was pretty typical for this class," says Bethencourt. "They're developing transgenic technologies that automate and greatly accelerate embryo transformation. It's likely to have a tremendous impact in pre-clinical medical innovation, given that their devices will speed current processes by 10 to 100 times. By the time they presented at Demo Day, they had $200,000 in presales."

Graduates from past classes also are making a mark, Bethencourt observes. Clara Foods, a Bay Area company that produces bioengineered, animal-free egg whites, has moved into series A funding.

"And Nerd Skincare [which produces bioengineered, antimicrobial skin-care products] has posted more than $1 million in sales," he says.

Bethencourt believes IndieBio is succeeding because its founders have adhered to their original mission: help scientists transform themselves into entrepreneurs.

"Scientists are no different from anyone else in terms of their basic motivations and desires," he says. "But because they've spent all their professional lives in academe and the lab, the business world can be very alien and intimidating to them. But they have to master that world if they're going to change the world as a whole. That's why the boot camp we've developed is so important. We push them out of their comfort zone. They have to learn accounting, sales, marketing, and PR. We make them conduct cold calls, talk to people with money, negotiate, and close deals."

Not everyone IndieBio considers—or even takes on—can make that leap, says Bethencourt.

"There's always an element of the unknown," he observes. "Some people you may have thought were naturals turn out incapable of adjusting to the demands of the marketplace. Others that you were sure were going to have trouble suddenly shine. One thing we've learned is to never judge companies until the end of the program. Or even *after* the program. Someone may struggle all the way through, and then all of a sudden figure it all out. We're here to support them until things clarify one way or the other—and we're always ready to be surprised."

So what's next for IndieBio?

"We think we have a template that can be applied internationally," he says. "We're working with people in Argentina and other countries. The United Nations contacted us, and we've been talking to their representatives. They're deeply invested in sustainable development, and they believe biotech is going to play a huge part in helping developing countries address their health care, food, and materials needs. We've come a long way in two years, but this is just the beginning. We're just getting started."

Glen Martin *covered science and the environment for the San Francisco Chronicle for 17 years. His work has appeared in more than 50 magazines and online outlets, including Discover, Science Digest, Wired, The Utne Reader, The Huffington Post, National Wildlife, Audubon, Outside, Men's Journal, and Reader's Digest. His latest book, Game Changer: Animal Rights and the Fate of Africa's Wildlife, was published in 2012 by the University of California Press.*

www.ingramcontent.com/pod-product-compliance
Lightning Source LLC
Chambersburg PA
CBHW060714030426
42337CB00017B/2870